XTREME AIRCRAFT

SPACEPLANES

A&D Xtreme
BOLD HI-LO NONFICTION
An imprint of Abdo Publishing
abdobooks.com

S.L. HAMILTON

TAKE IT TO THE XTREME!

GET READY FOR AN XTREME ADVENTURE!
THE PAGES OF THIS BOOK WILL TAKE YOU INTO THE THRILLING WORLD OF SPACEPLANES.
WHEN YOU HAVE FINISHED READING THIS BOOK, TAKE THE XTREME CHALLENGE ON PAGE 45 ABOUT WHAT YOU'VE LEARNED!

ABDOBOOKS.COM

Published by Abdo Publishing, a division of ABDO, PO Box 398166, Minneapolis, Minnesota 55439.

Printed in the United States of America, North Mankato, MN.

092021

012022

THIS BOOK CONTAINS RECYCLED MATERIALS

Editor: John Hamilton

Copy Editor: Tamara L. Britton

Graphic Design: Sue Hamilton

Cover Design: Laura Graphenteen

Cover Photo: Virgin Galactic

Interior Photos & Illustrations: CNSA-pg 36 (bottom inset); ESA-pgs 28-29 & 40-41; FAA-pg 43 (inset); Georges Méliès'-pg 7; Getty Images-pgs 36-37; ISRO-pgs 30 & 31; Molniya Research Industrial Corp-pgs 18-19; NASA-pgs 4-5, 8-9, 12-13, 14-15, 16-17, 20-21 & 38-39; Science Source-pgs 26-27; Scribner, Armstrong & Company-pg 6; Shutterstock-pg 1; Smithsonian Institution-pgs 32-33; US Air Force-pgs 10-11 & 24-25; Virgin Galactic-pgs 34-35, 42-43 & 44; XCOR Aerospace-pgs 22-23.

LIBRARY OF CONGRESS CONTROL NUMBER: 2021943546

PUBLISHER'S CATALOGING-IN-PUBLICATION DATA

Names: Hamilton, S.L., author.

Title: Spaceplanes / by S.L. Hamilton

Description: Minneapolis, Minnesota : Abdo Publishing, 2022 | Series: Xtreme aircraft | Includes online resources and index.

Identifiers: ISBN 9781532197369 (lib. bdg.) | ISBN 9781098219567 (ebook)

Subjects: LCSH: Aviation--Juvenile literature. | Aerospace planes--Juvenile literature. | Transatmospheric aircraft--Juvenile literature. | Space flight--Juvenile literature.

Classification: DDC 629.1333--dc23

TABLE OF CONTENTS

CHAPTER 1

SPACEPLANES

A spaceplane is able to travel within Earth's atmosphere, as well as in space. Brilliant scientists and engineers from around the world are making these **science fiction** ships a reality today.

Dream Chaser and other spaceplanes often have very similar shapes with boxy, lifting bodies and fixed wings.

CHAPTER 2

SPACEPLANE HISTORY

People have been wishing for spaceplanes for hundreds of years. Authors and artists were some of the first people to develop ideas for aircraft that could travel through Earth's atmosphere and into space.

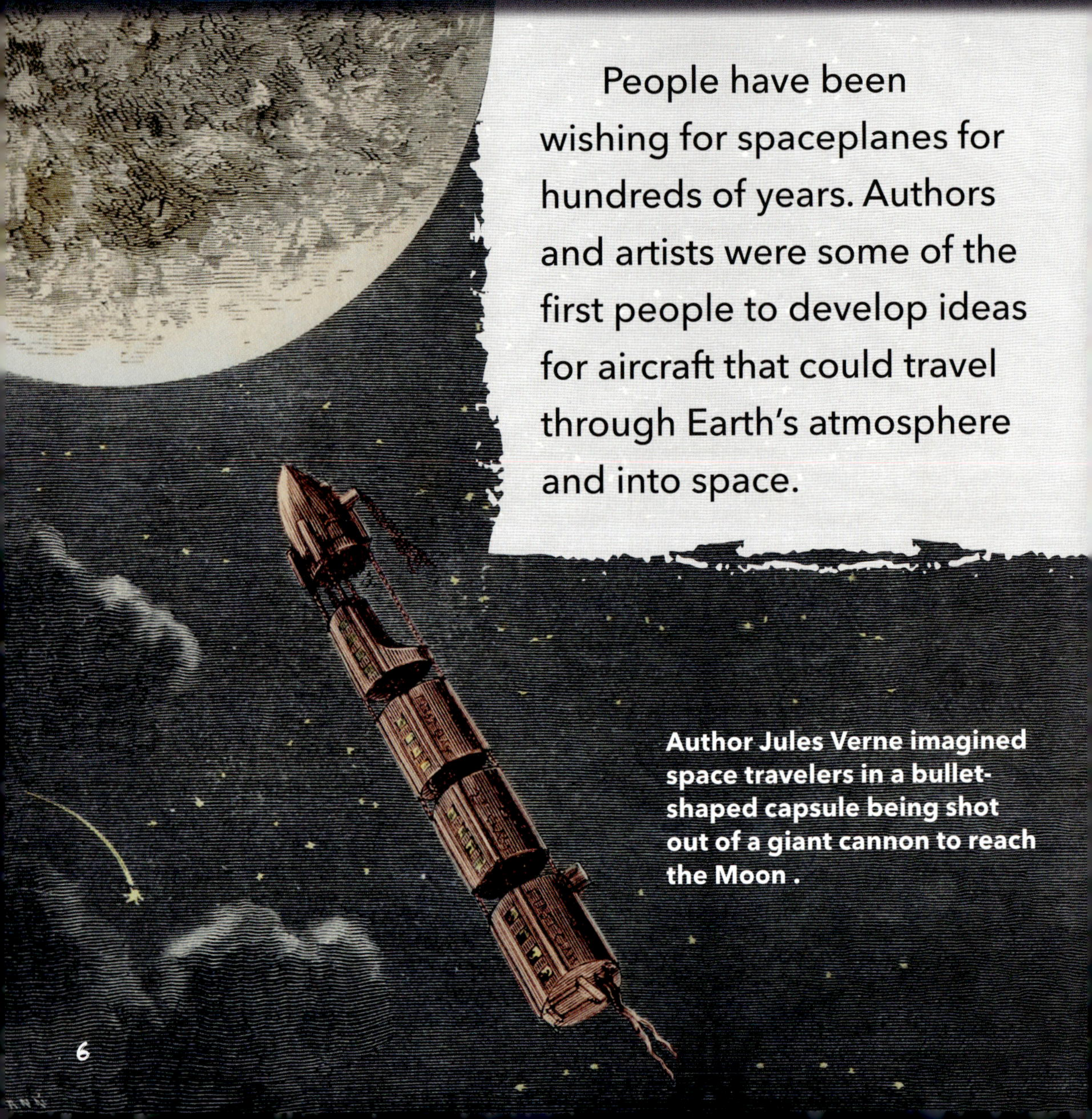

Author Jules Verne imagined space travelers in a bullet-shaped capsule being shot out of a giant cannon to reach the Moon .

More than 150 years ago, **science fiction** author Jules Verne wrote about round-trip space travel in *From the Earth to the Moon and a Trip Around It*. Filmmaker and artist Georges Méliès, inspired by Verne, created a short film, *A Trip to the Moon*, in 1902.

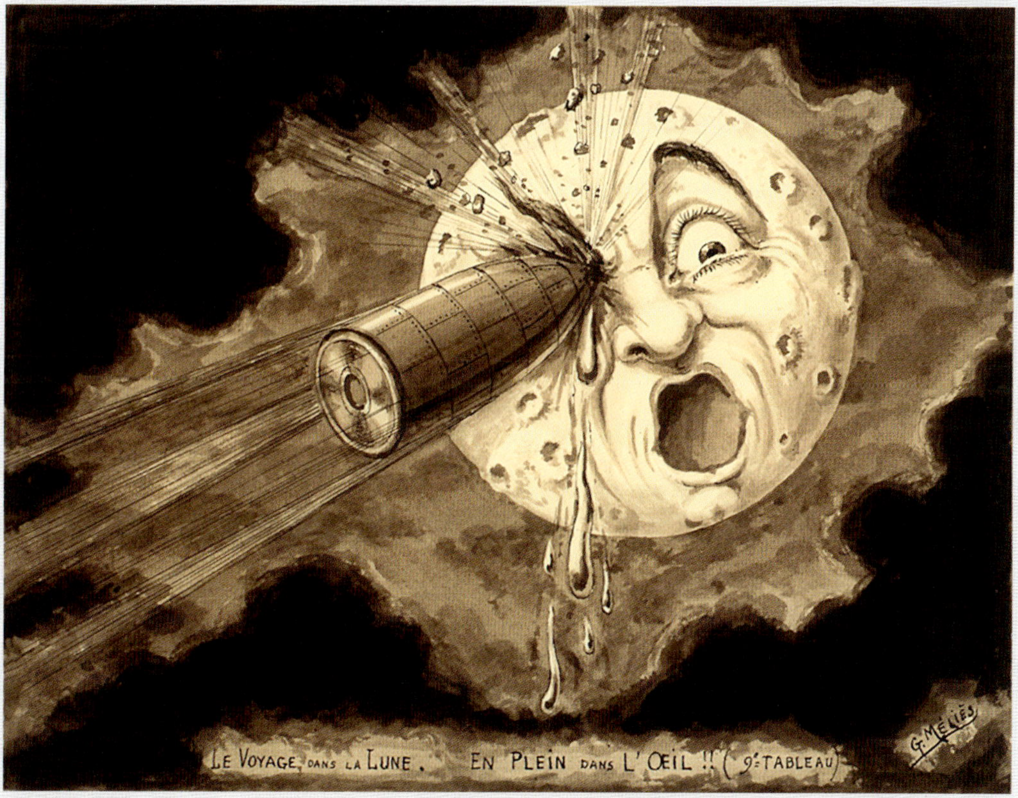

A spacecraft hits the Moon in a poster for the 1902 film *A Trip to the Moon*.

Design began for the first American-manned **rocketplane** in 1943. The XS-1, commonly known as the X-1, became the first aircraft to fly faster than the speed of sound.

Bell Aircraft built the first experimental X-1, as well as several more rocket-powered X-models.

Pilot Chuck Yeager flew the X-1 through the **sound barrier** on October 14, 1947. Going faster than **Mach 1** was an important speed accomplishment. From there, even faster planes were developed.

XTREME FACT

Two hours after it was achieved, Captain Yeager's supersonic flight was made top secret. The US Air Force did not officially confirm the feat until March 1948.

The US Air Force created the X-20 Dynamic Soarer in the 1950s and 1960s. The reusable, **supersonic** aircraft was designed to be launched using a rocket. It would then land on an airstrip under its own power. It was built and tested, but never actually flew as other space missions took priority.

The X-20's nickname was "Dyna-Soar."

XTREME FACT

Besides military uses, it was thought that the X-20 could be used for space rescue and satellite maintenance.

The North American X-15 was built to test aircraft structures, materials, and surfaces. Scientists needed to know how **rocketplanes** would work at **hypersonic** speeds and very high altitudes. The first of three X-15s rolled out on October 15, 1958.

XTREME FACT

The X-15 is the fastest manned aircraft ever flown. US Air Force Major Pete Knight went to Mach 6.72, or 4,520 mph (7,274 kph), on October 3, 1967.

An X-15 in flight.

An X-15 flight lasted about 10 minutes. During that time, only 85 seconds were fuel-powered flight. The rest of the time, the rocketplane was a glider. These aircraft helped NASA build the US space program and future spaceplanes.

Enterprise was NASA's first space shuttle. It rolled out on September 17, 1976. Two-man astronaut crews tested its structure, braking, controls, and free-flight landings within Earth's atmosphere. *Enterprise* did not have engines or working heat shields, so it could not go into space. The test flights helped scientists create future space shuttles.

Enterprise launches from NASA's 747 Shuttle Carrier Aircraft to begin a powerless glide test flight and landing.

XTREME FACT

Enterprise's **name came from the 1960s *Star Trek* TV series. Thousands of fans of the show wrote to President Gerald Ford. He asked NASA to use the popular TV spacecraft's name.**

CHAPTER 3

EARLY SPACEPLANES

NASA's shuttle program was officially called Space Transportation System (STS). *Columbia's* first flight was STS-1.

America's space shuttle program took off with *Columbia's* launch on April 12, 1981. Astronauts John Young and Robert Crippen flew the spaceplane on a two-day mission before landing it safely at Edwards Air Force Base. Future shuttles flew astronauts and **payloads** to and from the International Space Station (ISS). The shuttle program ended in 2011.

XTREME FACT

NASA built five space shuttles: *Columbia*, *Challenger*, *Atlantis*, *Discovery*, and *Endeavour*. Two shuttles and their crews were destroyed in accidents: *Challenger* (1986) and *Columbia* (2003).

Buran awaits launch in 1988. *Buran* is the Russian word for snowstorm or blizzard.

Buran was the **Soviet Union's** first spaceplane. Its only launch took place on November 15, 1988. The **unmanned** mission was a success. But money for the spaceplane's program ran out. Buran never flew again.

CHAPTER 4

MILLENNIUM MISSIONS

Lockheed Martin built NASA's X-33 Reusable Launch Vehicle (RLV) in 1999. It was expected to lower the cost of sending a payload into space from $10,000 per pound to $1,000 per pound. The robotic spaceplane was designed to reach speeds of more than **Mach 11**. NASA stopped the X-33 program in 2001.

A simulator is used to give scientists flight information on the X-33.

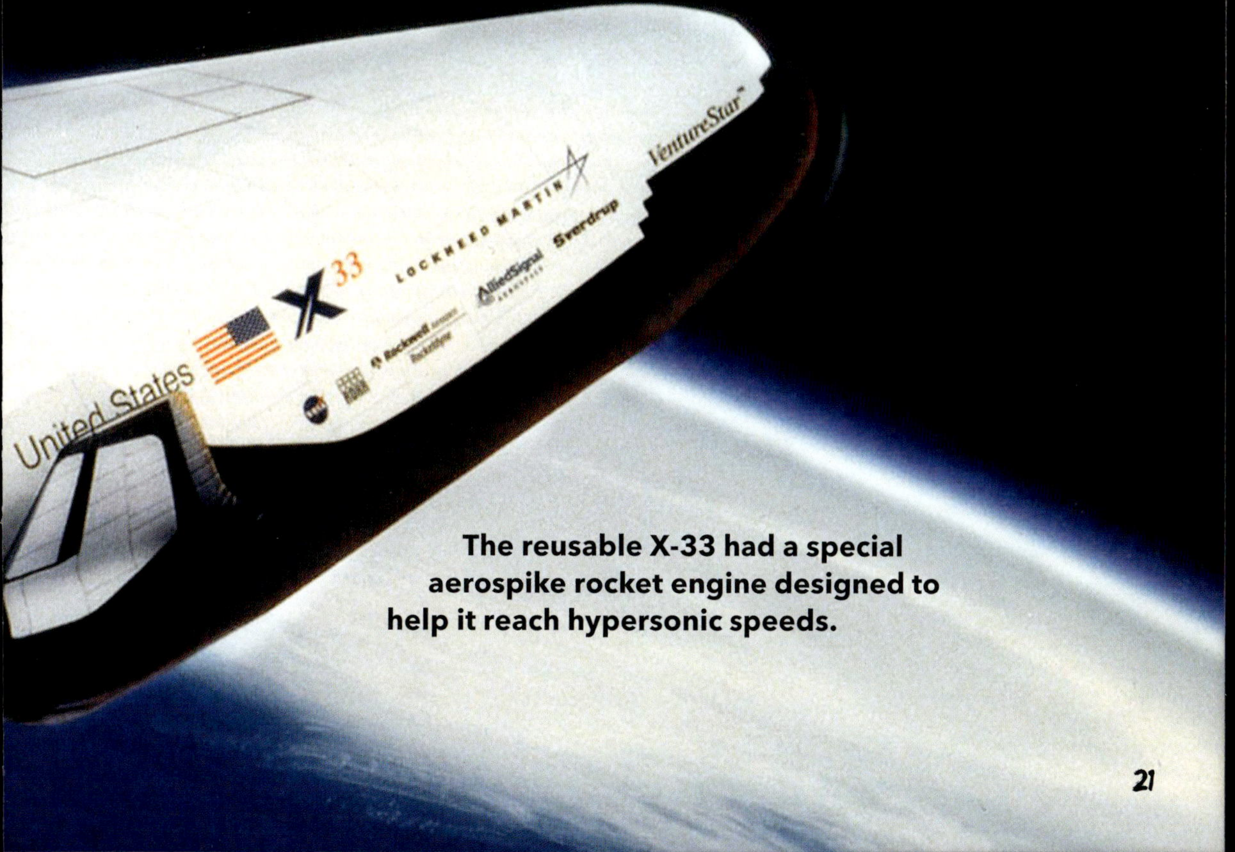

The reusable X-33 had a special aerospike rocket engine designed to help it reach hypersonic speeds.

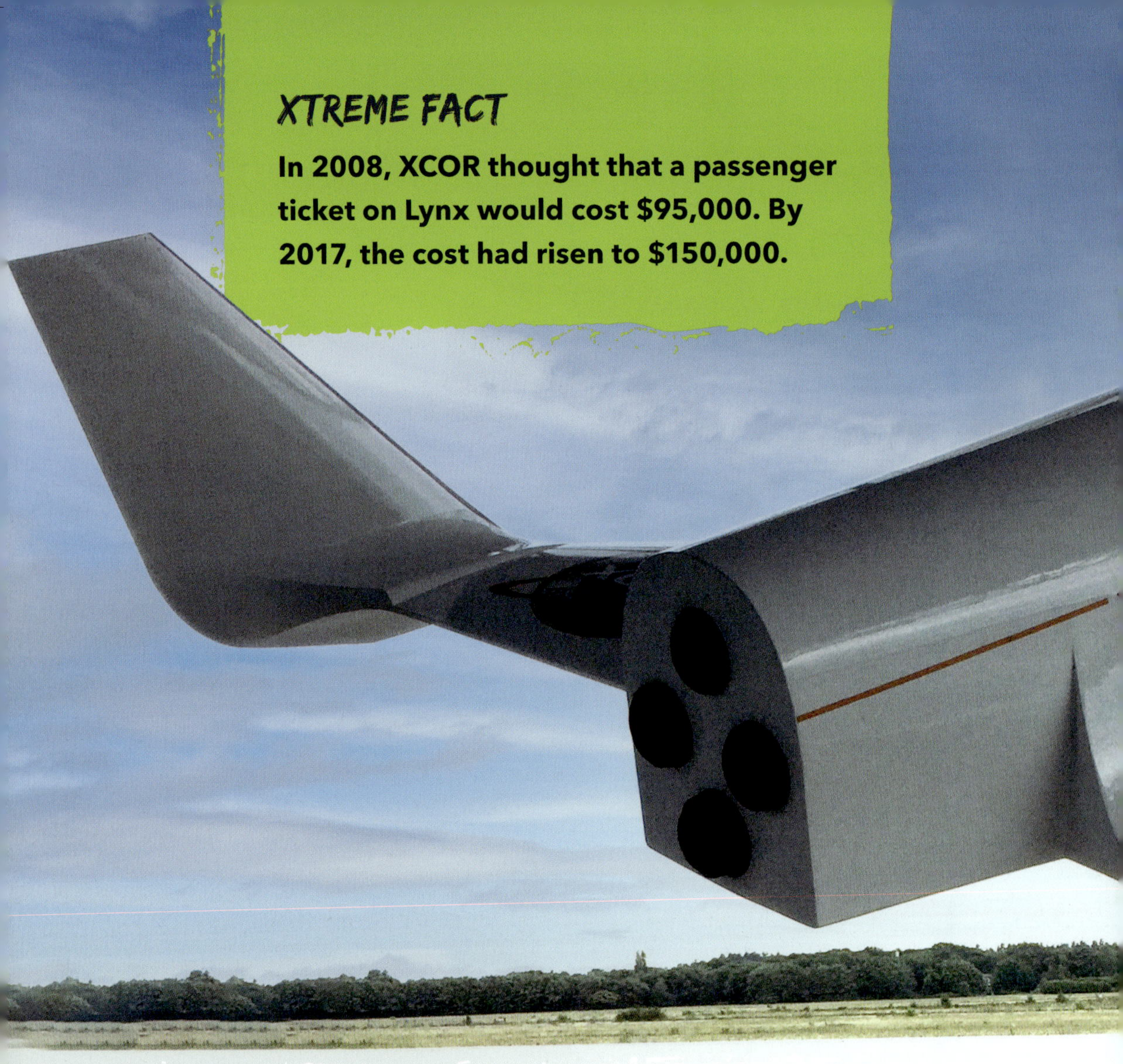

XTREME FACT

In 2008, XCOR thought that a passenger ticket on Lynx would cost $95,000. By 2017, the cost had risen to $150,000.

XCOR Aerospace's Lynx was a **horizontal-takeoff, horizontal-landing** (HTHL) spaceplane that began development in 2008. It was designed to carry a pilot, one passenger, and a **payload**. Unfortunately, problems caused the company to shut down in 2017. While a full-scale model was built, Lynx never flew.

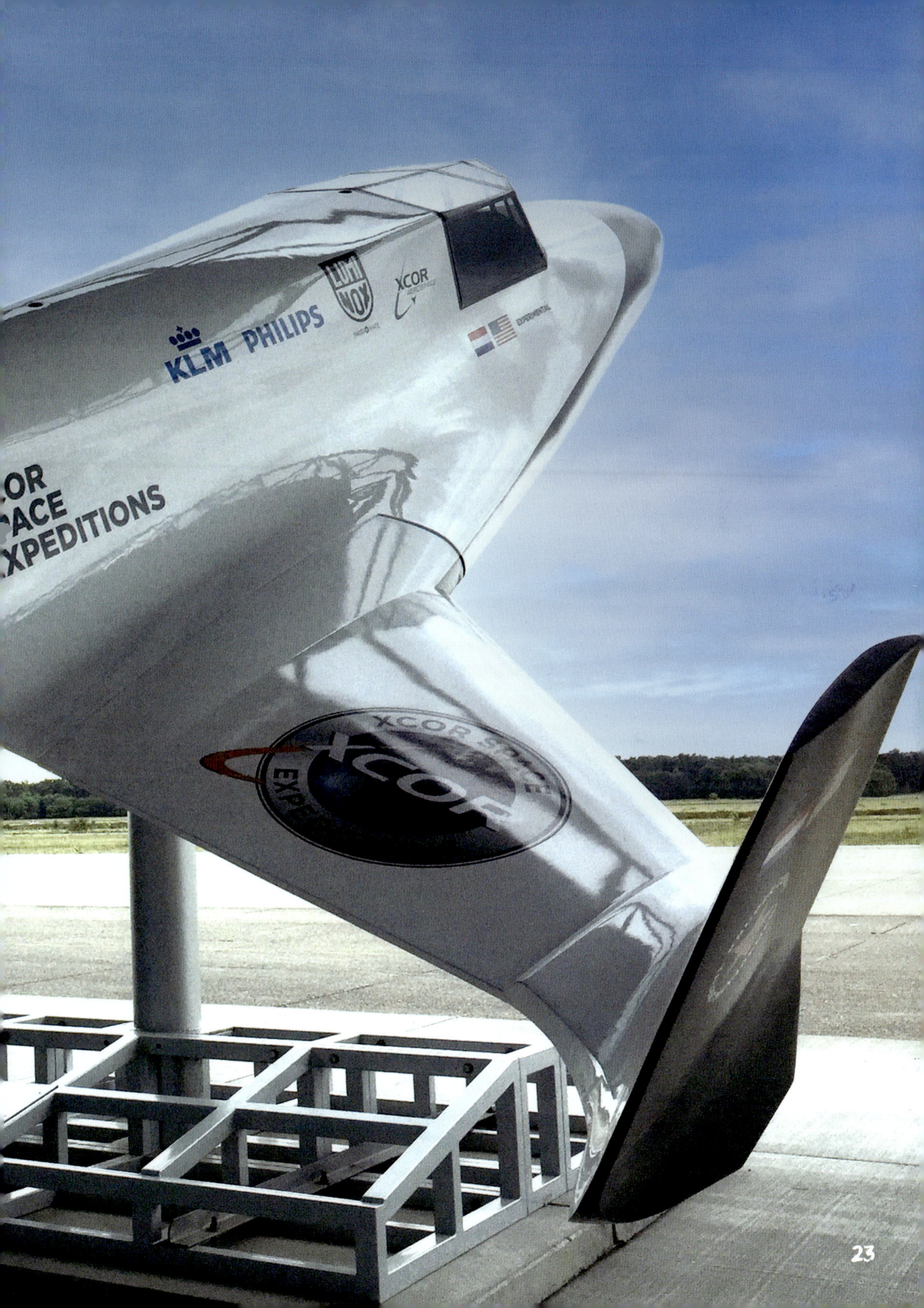
KLM
PHILIPS
XCOR
OR
ACE
EXPEDITIONS
XCOR SPACE
XCOR

CHAPTER 5

ROBOTIC SPACEPLANES

Boeing's X-37B Orbital Test Vehicle (OTV) was built for the US Air Force. The **unmanned** spacecraft launches vertically using an Atlas V or Falcon 9 rocket, but lands like a plane.

XTREME FACT

The X-37B broke its own in-orbit record, staying in space and conducting experiments for 780 days. It successfully returned to Earth on October 27, 2019.

The X-37B at NASA's Kennedy Space Center Shuttle Landing Facility in 2019.

The X-37B first flew on April 22, 2010. Ongoing missions are testing the OTV as a reusable craft that takes space experiments into orbit and returns them to Earth.

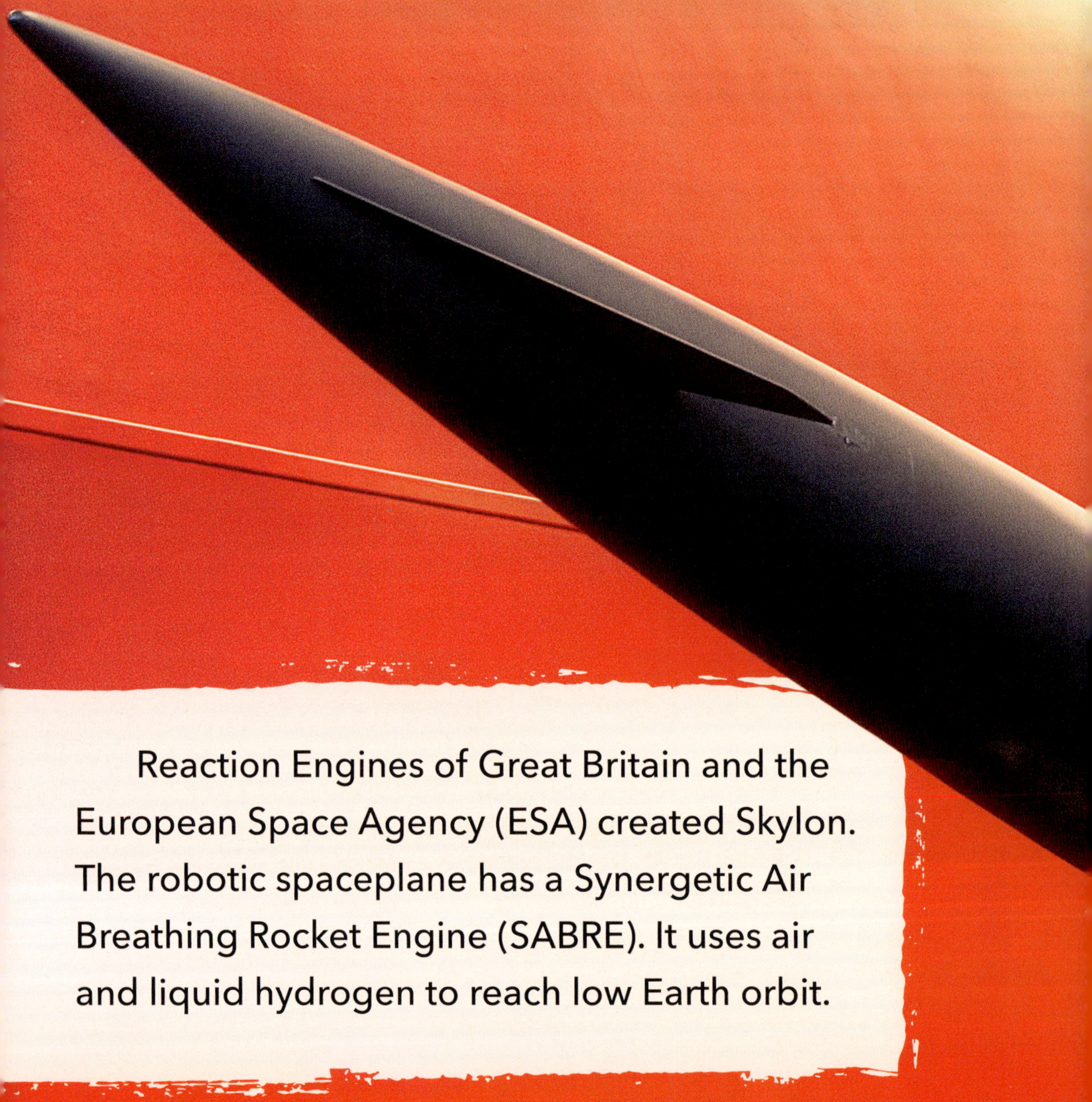

Reaction Engines of Great Britain and the European Space Agency (ESA) created Skylon. The robotic spaceplane has a Synergetic Air Breathing Rocket Engine (SABRE). It uses air and liquid hydrogen to reach low Earth orbit.

Skylon is a horizontal takeoff-horizontal landing spaceplane, making it able to use regular runways.

ESA's IXV reusable spaceplane is 16.4 feet (5 m) long. It's about the length of a midsize car. On February 11, 2015, it reached low Earth orbit, then parachuted for **splashdown** in the Pacific Ocean. The mission was designed to test reentry from space and lasted 1 hour and 40 minutes.

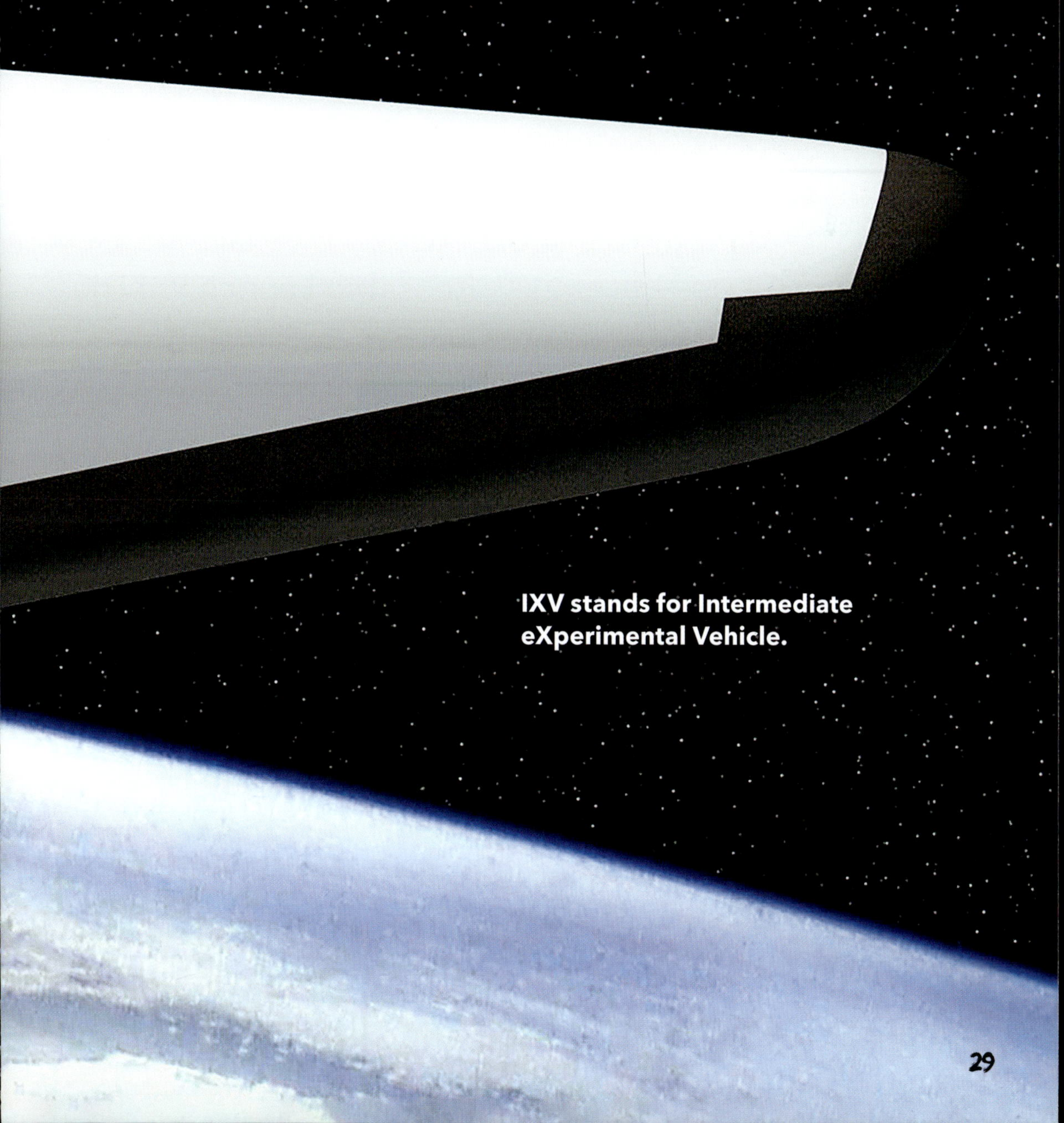

IXV stands for Intermediate eXperimental Vehicle.

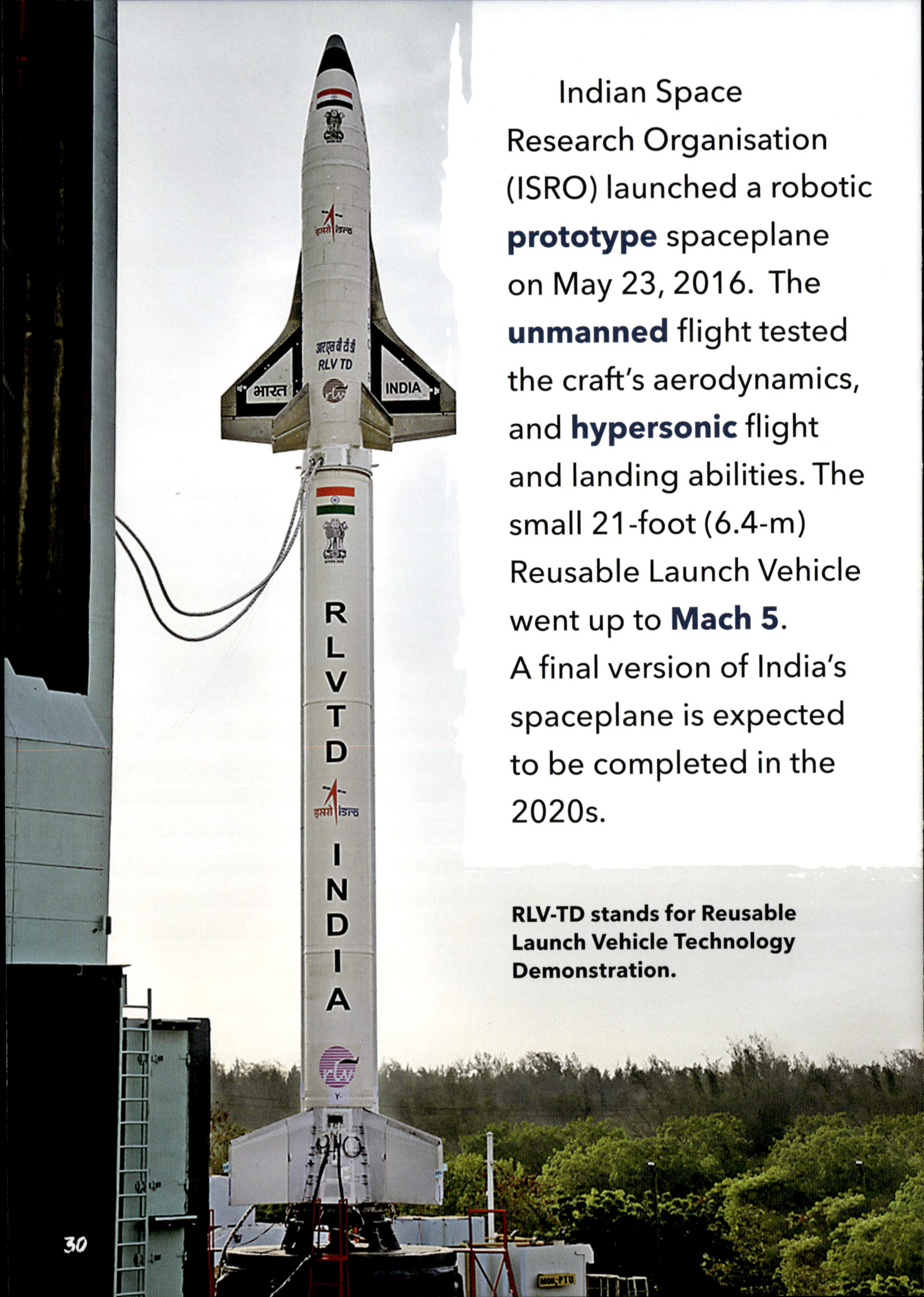

Indian Space Research Organisation (ISRO) launched a robotic **prototype** spaceplane on May 23, 2016. The **unmanned** flight tested the craft's aerodynamics, and **hypersonic** flight and landing abilities. The small 21-foot (6.4-m) Reusable Launch Vehicle went up to **Mach 5**. A final version of India's spaceplane is expected to be completed in the 2020s.

RLV-TD stands for Reusable Launch Vehicle Technology Demonstration.

ISRO's Reusable Launch Vehicle's successful test flight lasted 770 seconds, or just under 13 minutes.

CHAPTER 6

SPACE TOURISM

SpaceShipOne was the first privately built aircraft to reach supersonic flight. It is now in the National Air and Space Museum in Washington, DC.

SpaceShipOne was built and tested in the early 2000s by Scaled Composites. It was developed as a spaceplane for tourist travel. Pilot Brian Binnie flew it on December 17, 2003, exactly 100 years after the Wright brothers flew the first powered aircraft in North Carolina.

SpaceShipOne's cockpit.

SpaceShipTwo, built for Virgin Galactic, is the next generation of spaceplane. It launches from a carrier aircraft at 50,000 feet (15,240 m). SpaceShipTwo's rocket engines reach **supersonic** speed in 8 seconds, lifting two pilots and six passengers into low Earth orbit.

Carrier aircraft WhiteKnightTwo flies with SpaceShipTwo (center).

Virgin Galactic's *Unity* flies free. The company plans to operate a fleet of five SpaceShipTwo spaceplanes from its New Mexico spaceport.

XTREME FACT

Virgin Galactic began space tourism flights on July 11, 2021. Founder Richard Branson and 5 others made a successful 53-mile (85-km) trip to the edge of space aboard *Unity*.

CHAPTER 7

MODERN MISSIONS

China has been testing several spaceplane designs. It launched a reusable spaceplane on September 4, 2020. The China Aerospace Science and Industry Corp (CASIC) is also working on a spaceplane. CASIC's Tengyun spaceplane could take off and land from an ordinary airport.

China's reusable spaceplane had a two-day mission before returning to Earth in September 2020.

A Long March 2F carrier rocket blasts off from China's Jiuquan Satellite Launch Center in the Gobi Desert. The same rocket and location were used to launch a reusable spaceplane in 2020.

One version of Sierra Nevada Corporation's Dream Chaser is a seven-crew spaceplane built for NASA. Another version is an **unmanned** cargo carrier designed to take supplies to the ISS. This resupply spacecraft is called *Tenacity*. Its planned launch date is in 2021.

Dream Chaser is an SUV: Space Utility Vehicle.

XTREME FACT

Dream Chaser is 30 feet (9 m) long. It would take about four of them to add up to the length of a NASA space shuttle.

ESA's Space Rider is being prepared for launch in 2022. The robotic craft is a high-tech space lab. It can perform experiments in low Earth orbit for more than two months. When it returns to Earth, a controllable gliding parachute called a parafoil will assist with the descent.

Space Rider stands for Space Reusable Integrated Demonstrator for Europe Return.

Space Rider will go up in a Vega-C carrier rocket. The robotic craft will perform experiments. When it returns to Earth, it will be picked up at a landing area. Data and experiments will be turned over to scientists. Then Space Rider will start all over again on new projects.

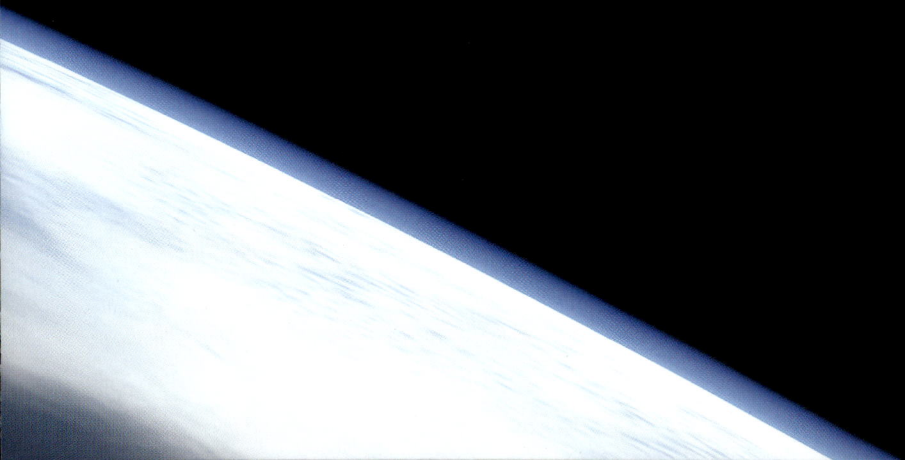

CHAPTER 8

BECOMING A SPACEPLANE PILOT

Spaceplane pilots are often former astronauts or test pilots. They spend thousands of hours training in experimental aircraft and flight simulators. Like anyone commanding an aircraft, pilots need to stay calm in emergencies and be ready for anything.

XTREME FACT

There is no official license for a spaceplane pilot, but America's Office of Commercial Space Transportation issues Astronaut Wings. Only seven people have received wings to date.

Virgin Galactic's Dave Mackay, Mike "Sooch" Masucci, and Beth Moses all received Astronaut Wings in 2019.

CHAPTER 9

THE FUTURE

Spaceplanes will soar into the future. Government-run programs such as those by NASA, ESA, ISRO, and CASIC continue to lead the way. But commercial companies such as Virgin Galactic are not far behind. Space tourism may begin in the 2020s.

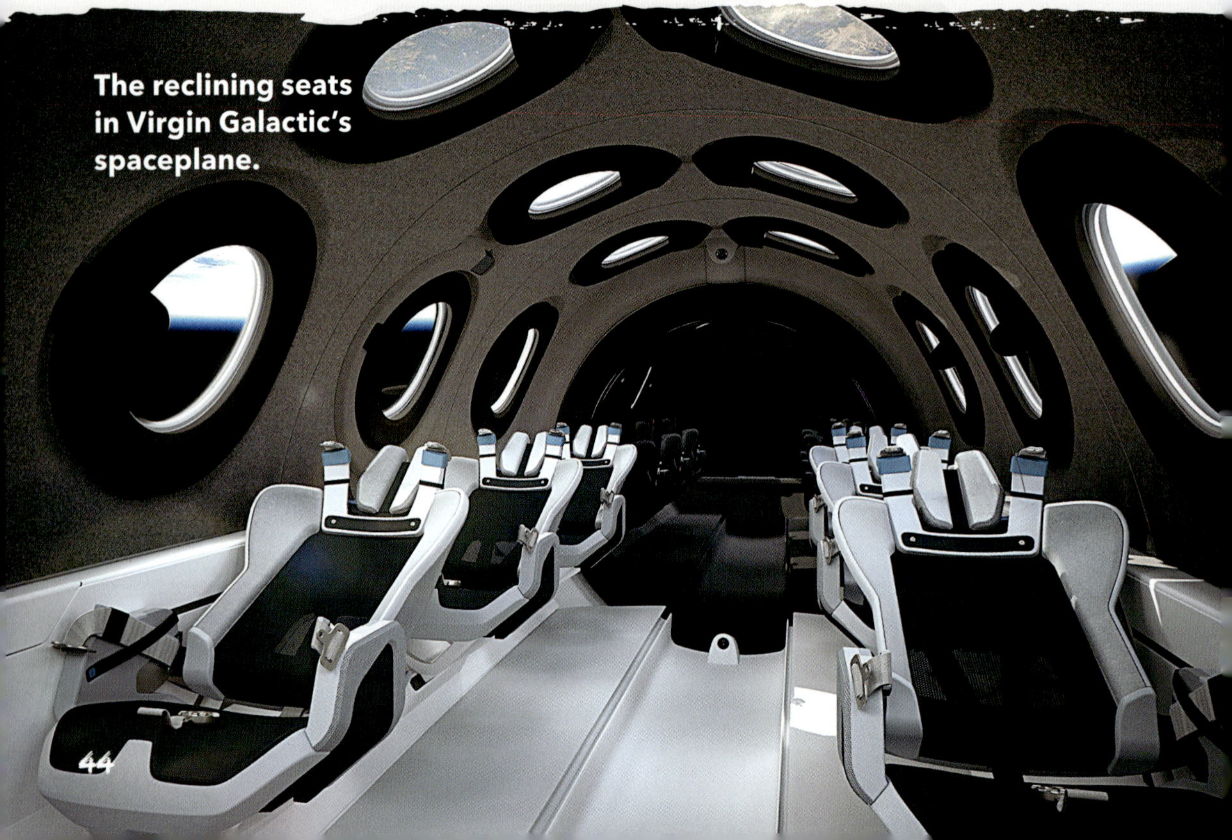

The reclining seats in Virgin Galactic's spaceplane.

XTREME CHALLENGE

TAKE THE QUIZ BELOW AND PUT WHAT YOU'VE LEARNED TO THE TEST!

1) What can a spaceplane do that other planes cannot?

2) Who was the first pilot to break the sound barrier? What experimental rocketplane did he use?

3) What was the name of the first space shuttle to go into space? What year did it go up?

4) SpaceShipOne was the first privately built spaceplane designed for tourist travel. Who was the first to fly it?

5) After launching from a carrier aircraft, how quickly can SpaceShipTwo's engines reach supersonic speed?

6) What is the name of the ESA spaceplane that is a "high-tech space lab"? Is it manned or unmanned?

7) How does someone become a spaceplane pilot? What important traits are needed?

GLOSSARY

horizontal take-off, horizontal landing – The ability of a spaceplane to take off and land from a runway, much like a regular plane.

hypersonic – Extremely high speed between 3,821-7,643 mph (6,150-12,300 kph).

Mach 1 – The speed of sound waves traveling through air at about 700 mph (1,127 kph) at 43,000 feet (13,106 m), the approximate altitude in which test pilot Chuck Yeager made his historic 1947 flight. Mach 2 is twice the speed of sound, Mach 3 is three times, and so on.

payload – Something that is carried by an aircraft, rocket, or missile. Payload may also refer to the weight of goods and people in a vehicle.

prototype – The first test model built of something new, such as an aircraft.

rocketplane – An aircraft that is powered by rockets.

science fiction – A type of fiction that involves the future and advancements in science. This may include stories about space travel, flying with jetpacks, or robots.

sound barrier – When a plane goes faster than the speed of sound (Mach 1), it is said to have broken the sound barrier.

Soviet Union – A former country that included a union of Russia and several other Communist republics. It was formed in 1922 and existed until 1991.

splashdown – When a spacecraft lands in water instead of on dry land.

supersonic – Speeds greater than the speed of sound. The range may go from about 700-3,820 mph (1,127-6,148 kph). Speeds above that range are called hypersonic.

unmanned – Without a human pilot or crew on board. Many spaceplanes are unmanned. Controls are manned by people on the ground or by robots.

ONLINE RESOURCES

Booklinks
NONFICTION NETWORK
FREE! ONLINE NONFICTION RESOURCES

To learn more about spaceplanes, please visit **abdobooklinks.com** or scan this QR code. These links are routinely monitored and updated to provide the most current information available.

INDEX